ANIMAL BABIES

Baby Moose

by Jennifer Boothroyd

Consultant: Beth Gambro
Reading Specialist, Yorkville, Illinois

Minneapolis, Minnesota

Teaching Tips

Before Reading

- Look at the cover of the book. Discuss the picture and the title.
- Ask readers to brainstorm a list of what they already know about moose. What can they expect to see in the book?
- Go on a picture walk, looking through the pictures to discuss vocabulary and make predictions about the text.

During Reading

- Read for purpose. Encourage readers to think about moose and the animal life cycle as they are reading.
- If readers encounter an unknown word, ask them to look at the sounds in the word. Then, ask them to look at the rest of the page. Are there any clues to help them understand?

After Reading

- Encourage readers to pick a buddy and reread the book together.
- Ask readers to name three things moose do between the time they are born and the time they are ready to have babies. Go back and find the pages that tell about these things.
- Ask readers to write or draw something that they learned about moose.

Credits:

Cover, © johan10/ iStock; 3, © Amanda Wayne/ Shutterstock; 5, © Design Pics Inc / Alamy; 6, © Design Pics Inc / Alamy; 7, © mlharing/ iStock; 7, © Early Spring/ Shutterstock; 8,9, © Jill Ann Spaulding/ Getty; 10, © Wichai Prasomsri1/ Shutterstock; 11, © Rocky Grimes/ Alamy; 12, © Wichai Prasomsri1/ Shutterstock; 13, © pchoui/ iStock; 15, © Tom Walker/ Alamy; 16,17, © Freder/ iStock; 17, © Wichai Prasomsri1/ Shutterstock; 18, © Kichigin/ Shutterstock; 18,19, © Mats Lindberg/ iStock; 20, © Wichai Prasomsri1/ Shutterstock; 20,21, © RichLegg/ iStock; 22, © Teemu Tretjakov/ Shutterstock; 23, © Wildnerdpix/ Shutterstock; 23, © Szczepan Klejbuk/ Shutterstock; 23, © Classic Style/ Shutterstock; 23, © Irina Markova/ Shutterstock; 23, © Design Pics Inc/ Alamy.

Library of Congress Cataloging-in-Publication Data

Names: Boothroyd, Jennifer, 1972- author.
Title: Baby moose / by Jennifer Boothroyd.
Description: Bearcub books. | Minneapolis, Minnesota : Bearport Publishing
  Company, [2022] | Series: Animal babies | Includes bibliographical
  references and index.
Identifiers: LCCN 2021026684 (print) | LCCN 2021026685 (ebook) | ISBN
  9781636913575 (library binding) | ISBN 9781636913643 (paperback) | ISBN
  9781636913711 (ebook)
Subjects: LCSH: Moose--Infancy--Juvenile literature.
Classification: LCC QL737.U55 B63725 2022 (print) | LCC QL737.U55 (ebook)
  | DDC 599.65/71392--dc23
LC record available at https://lccn.loc.gov/2021026684
LC ebook record available at https://lccn.loc.gov/2021026685

Copyright © 2022 Bearport Publishing Company. All rights reserved. No part of this publication may be reproduced in whole or in part, stored in any retrieval system, or transmitted in any form or by any means, electronic, mechanical, photocopying, recording, or otherwise, without written permission from the publisher.

For more information, write to Bearport Publishing, 5357 Penn Avenue South, Minneapolis, MN 55419. Printed in the United States of America.

Contents

It's a Baby Moose! 4

The Baby's Body 22

Glossary 23

Index 24

Read More 24

Learn More Online......................... 24

About the Author 24

It's a Baby Moose!

A baby moose tries to stand.

It **wobbles** on long legs.

Then, it falls down.

Plop!

This moose **calf** is as tall as a kitchen table.

The baby has brown fur.

It has a very short tail.

The calf drinks milk from its mother's body.

Then, the baby falls asleep.

Baby moose take naps every day.

After its nap, the calf tries standing again.

It can stand for longer.

The calf walks near its mother.

Soon, it can run and jump.

The calf grows bigger every day.

In a few weeks, it learns to find food.

It starts to eat grass and leaves.

The calf also eats plants that grow in water.

So, the baby moose learns how to swim!

It holds its breath **underwater**.

After a few months, a boy calf grows **antlers**.

They begin as hard bumps on its head.

The antlers will grow bigger each year.

Soon, it is winter.

The grass is covered with snow.

The calf and its mother eat **twigs**.

When spring comes, the young moose leaves its mother.

It lives alone.

In a year or two, it can have its own baby.

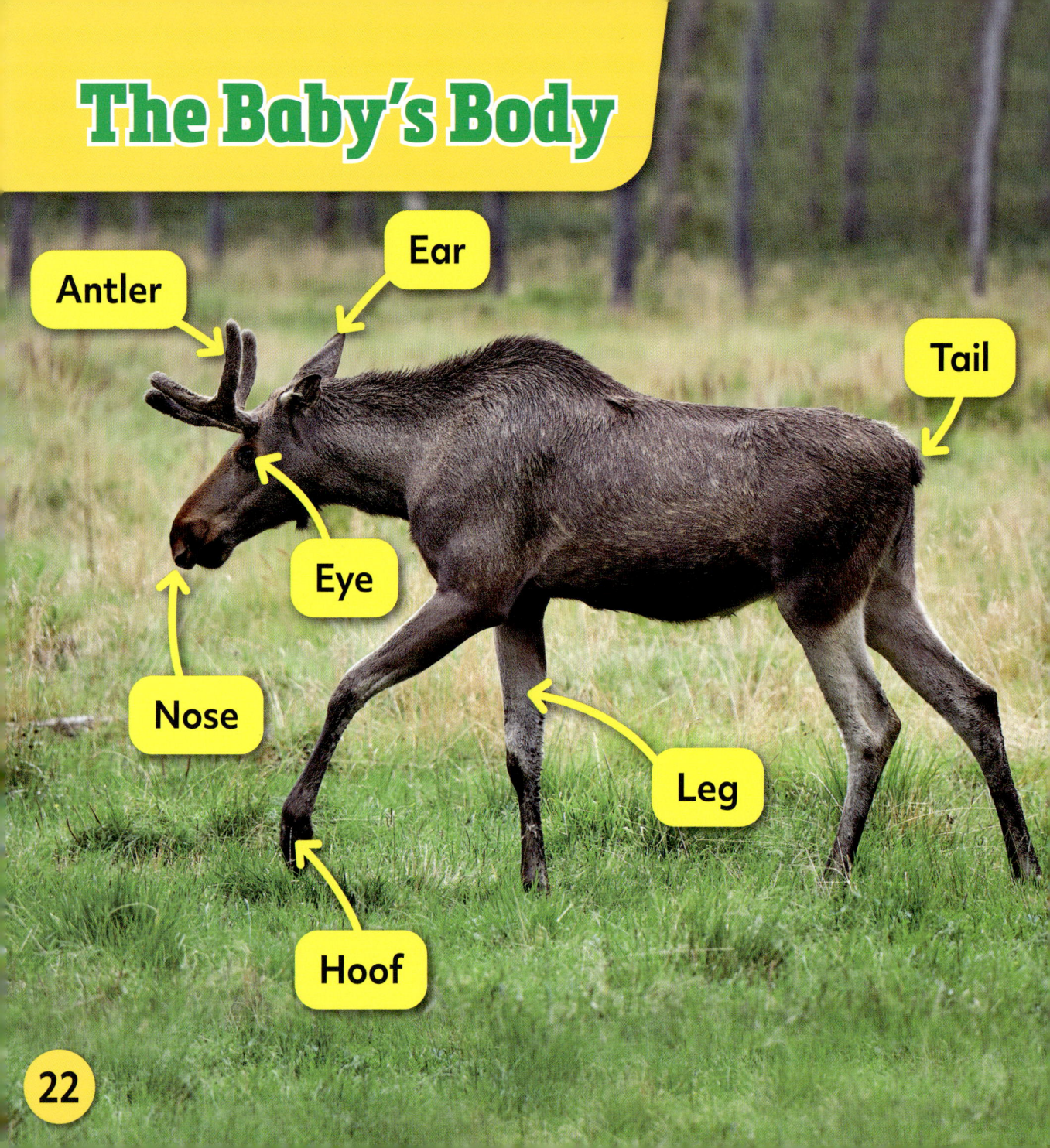

Glossary

antlers bony parts that grow on some animals' heads

calf a baby moose

twigs little branches of trees or bushes

underwater under the water

wobbles shakes

Index

antlers 16–17, 22
eat 12, 14, 18
fur 7
milk 8
mother 8, 10, 18, 20
plants 14
swim 14

Read More

Albertson, Al. *Moose (Blastoff! Readers: Animals of the Forest).* Minneapolis: Bellwether Media, 2020.

Loh-Hagan, Virginia. *Moose (My Favorite Animal).* Ann Arbor, MI: Cherry Lake Publishing, 2018.

Learn More Online

1. Go to **www.factsurfer.com** or scan the QR code below.
2. Enter "**Baby Moose**" into the search box.
3. Click on the cover of this book to see a list of websites.

About the Author

Jennifer Boothroyd lives in northern Minnesota. Moose live there, too.